The COUNTRY STORE
The General Store of Yesterday

By ELMER L. SMITH

Photography by MEL HORST

Third Printing 1982

Applied Arts Publishers
Lebanon, Pennsylvania 17042

ISBN: 0-911410-43-0

The Country Store's Forerunners

In the early days of our nation's settlement there was a vast difference in the standard of living, culture and level of civilization between eastern sea-coast towns and remote inland regions. Commerce developed gradually but steadily along the coast, but inland areas were sparsely peopled by trappers and traders and pioneer settlers, all facing primitive conditions and whose major energy was expended for their very survival.

The Peddler Serves the Pioneer

Early peddlers linked these two worlds, offering needed goods and news from the outside world. The first peddler traveled by foot with a backpack or trunk, although some rode horseback with saddle bags jammed with merchandise. The very nature of the travel limited his products to small necessities—there were pins, needles, spoons, knives, buttons, spices, combs and other small objects, although costly for the size. Much trade was by barter because there were few dollars available in the self-sufficient farmsteads, nor were banks yet established. Items were sold for trade goods which also had to be limited to those items easily portable, such as fur pelts.

The peddler has been characterized as an unstable, vagrant type who was shrewd but lacking in ambition—but it must be noted that this stereotype is misleading—some of our great corporations emerged from the efforts of a peddler, and some leading inland stores were established by peddlers who settled down.

Travel conditions were so poor and body comforts so limited that it is hard to determine what motivated these early peddlers. But there were enough of such itinerants traveling the inland regions to create a special type of commerce appropriate to the time, place and conditions. It was these unique individuals who became the inland source of world goods which they bought direct at port cities from importers or shippers.

Samuel Alexander & Charles B. Penrose

Have bought from George Gaullagher his large and very extensive stock of

Merchandize,

composing a most extensive assortment of GOODS, suitable for the present and approaching seasons, and which, (at the old stand of *George Gaullagher*,) they now offer to their friends and the public, at the

Most Reduced Prices.

From a determination to keep the assortment at all times *full*, and their disposition to accommodate all who may favour the store with a call, they declare that on their part, nothing shall be wanting to afford satisfaction.

The following articles compose a part of their *STOCK OF GOODS*, to wit:

Angola Cassimeres,
Plain and striped Satinetts,
Bombazets and Bombazeens,
Irish Popolins,
Striped Bengals,
Blue and yellow Company Nankeens.
Levantine, Senshaws, Mantuas, Florence and Sarsnett Silks,
Plain and figured Mull Mull,
Jaconet, Cambrick and Swiss Muslins,
Robinets and Italian Crapes,
Bengal Chintz and Ginghams,
Long Lawn and Linen Cambricks,
Washington, Wilmington & Union Stripes
Painted Muslins and Bed Ticking,
Wash Leather, Horse & Dog skin Gloves,
Silk, Kid, and York tan Gloves,
Gentlemen and Lady's Leghorn Hats,
Straw and Gimp Bonnets,

ALSO,

Rock and Rifle Powder, Brandy, Gin, Spirits, Molasses, Sugar, Coffee, Tea, Pepper, Alspice, Salt, Fish, &c. &c.

August 17, 1825.

Gradually the pack peddlers became wagon peddlers, a trend obvious by 1800 with the wave of road building projects throughout the new republic. The first turnpike was initiated in 1785, running from Alexandria into the lower Shenandoah Valley of Virginia; in 1787 the Baltimore-Frederick-York road; in 1795 the Philadelphia-Lancaster turnpike and others such as the Boston Post road. These were toll roads, but together they served only a small portion of the pioneer and early inland settlements. Road building to serve all the people came much later and the peddling commerce flourished! Estimates run from 10,000 to 15,000 peddlers in 1850 and larger numbers by 1860 as the population increased and inland areas became more accessible for homesteads.

Villages and Towns

With the improved roads and the established routes, inns and general merchandise stores became a part of the progress,

Early engraving of a peddler displaying his wares. An 1825 full-si hand-bill lists merchandise available to wholesalers.

Kauffman's in New Holland, Pennsylvania, has been in continuous business since 1779, originally serving as a general store and tavern house. The photograph (above) was taken about 1905. It is a truly historic sight having served at various times as post office, town hall, home of the early fire company and as a school, meeting place for quiltings, oyster parties and even the Anti-Masonic party.

and as towns and villages emerged along these roads, the general store was created to serve the needs of the people. Obviously, there was competition between the locally established enterprise and the itinerant. Store operators pressured governments to place controls on the itinerant and some states adopted licensing and other restrictive measures. The term "damn Yankee" characterized the New England peddlers who were "slick, if not downright crooked traders."

One of the country store's competitors was organized mass peddling such as established by J. R. Watkins with his "*store at your door*" system of marketing. Watkins's was a Minnesota medicine-maker who sold his products door-to-door from his horse-drawn wagon. He had a simple technique based on honesty—never claim a "sure cure" for anything as a result of using the Watkins product.

As Watkins expanded his line, other peddlers took on his products and by his fifteenth year (1880), he had over two hundred items including remedies, flavorings, polish, beauty aids, cosmetics, candy and even vehicle parts! Hundreds of Watkins wagons were on the roads reaching across the growing nation with his products.

Door-to-door peddling evolved into the chain store—the general store's most fierce competitor. The chain store had the advantage of mass buying, professional management and pre-packaging—perhaps a leading example was the Great Atlantic and Pacific Tea Company.

In 1859 George Hartford, a youth from Maine, noted that tea was bought at the docks by importers who sold it to wholesalers and peddlers-retailers with the result that it had three mark-ups before it reached the family table. He bought an entire cargo and sold it at half the typical price and still made a profit. Later he opened a retail store in New York City, painting it vermilion red and gold to attract attention. His products expanded to coffee, extracts and flavorings and when the transcontinental railroad was completed in 1869, he named his company The Great Atlantic and Pacific Tea Company. By 1880 he had 95 stores from Boston down the east coast and west to Milwaukee. To serve the rural people, he established a fleet of red and gold horse-drawn wagons to peddle A & P products direct to the home.

Impact of Chain-Store and Mail-Order

Beginning in 1910, the mobile stores were replaced with stores in smaller towns making the A & P America's largest chain grocery company; but more significantly, it perhaps gives the best evidence of the beginning of the decline of the general stores. This period coincides with the rise of the Motor Age which facilitated private transportation to the extent that the vast proportion of the population, even in remote areas, could travel to nearly towns.

Another competitor of the local general store emerged as the mail-order house, such as Sears Roebuck & Co., estab-

Continued on page 5

Barrels, Boxes, and Bins

In early times when a customer came to the general store, they didn't have a choice of brands—they simply asked for a "slab of bacon," a "wedge of cheese," a "bag of flour" and the storekeeper weighed out and wrapped each item and placed a penciled price on it.

The merchant received commodities in bulk—barrels of flour, boxes of crackers, crates of fruits, firkins of lard, bags of cereals, dried beans and sugar. Coffee was in green bean form and had to be weighed, ground and roasted at the store or in the home. Nutmeg, pepper, cloves were in peppercorn state and were ground at home on a grater or in a special grinder. Vinegar was pumped directly from the barrel, molasses was fauceted directly from the shipping hogshead, as was kerosene and other liquids. Tea was sold loose from a large counter canister. Peas, grains, beans, rice and similar items were in large bins or built-in drawers. Cheese came in large wheels, the clerk cutting wedges to order.

Thus each sale was a custom order and every item was measured, weighed out by the clerk, wrapped and tied with a string and priced. The completed order was placed in the customer's market basket. (Later, after the development of the paper bag, the clerk usually itemized and totaled the order on the bag.)

Bulk merchandising raised concern about sanitary conditions and accuracy of the scales and measurements. Folk stories are told about some storekeepers

The large cheese wheel was ordinarily placed on the cheese cutter (below) which cut entirely through to the center point. The wooden pump was hand operated and could reach the lowest levels of the barrel. Mince Meat, candy and other products were often packed in wooden pails which served as the store containers.

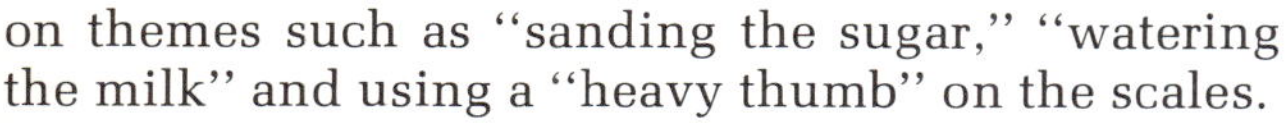

on themes such as "sanding the sugar," "watering the milk" and using a "heavy thumb" on the scales.

End of The Cracker Barrel

The old cracker barrel days ended when the National Biscuit Company packed its famous *Uneeda* Soda Crackers in a family-size moisture-proof pack in 1898. The cheese wheels declined after Kraft developed processed cheese packaged in small, family-sized containers.

Widespread change took place beginning in the 1860's and at the turn of the century, products came to the general store individually wrapped in tin containers, paper boxes, sealed bags and in bottles, carrying bright colored eye-appealing labels.

The Country Store's Forerunners

Continued from page 3

lished by Richard W. Sears, a telegrapher and station agent for a railroad in Spring Valley, Minnesota. When a shipment of watches was refused, he bought them and sold them to railroaders. He repeated the process and later formed the R. W. Sears Watch Company. When Sears needed the services of a watch repair man, he advertised and selected an applicant, Alvah C. Roebuck. Together they expanded the product line into general merchandise.

The mail-order establishments issued descriptive catalogues of an assortment of items far in excess of that which would be available from a general store or a peddler without any problems with credit and delivery. The mail-order catalogue became a major "literary" publication, looked forward to by country folks as much as any monthly magazine and probably read more than most family Bibles! The mail-order catalogues became known as the "Dream Book," and it was often used for comparative shopping—keeping the local store in line price-wise!

Exterior view of Kreider's store, Witmer, Penna., as it appeared in the winter of 1909.

Below are shown the grocery section of Kreider's Country Store in 1907 and the clothing and dry goods departments seventy years ago.

Miller's General Merchandise—1851 to Now

In 1851 when Joel Miller opened a "General Merchandise" store in the small village of Witmer, Lancaster County, Pennsylvania, it was a community of a few scattered farms and narrow dirt roads for horse and wagon. Families "went to town" seldom and even a trip to the crossroads store was a once-a-week occasion. During the 46 years Miller operated the store, dramatic changes were taking place throughout America but the primary impact was in the rising cities rather than the quiet, self-sufficient farms of the thrifty Pennsylvania Germans.

In 1897, the store was sold to Harry Kreider who continued the sale of clothing, hardware, notions and food, but the turn of the century was a different era. Brand names were widely introduced, national advertising was on the increase, and pre-packaged goods now dominated the shelves.

This sequence of early photographs portray the Miller store with its fixtures, equipment and merchandise. Open barrels, boxes and tins denote that goods were still available in bulk.

Survived until 1928

After a quarter century, at Kreider's death, the store was leased to Harry Heller, but its operation was short-lived and the doors were closed in 1928. (Most experts mark 1930 as a period of transition when the general stores of America faced the final

test of survival—and most of them failed!) Few could compete with the new trends—the rise of the specialty store, the improved roads and wide-spread use of the automobile, increased competition from national grocery chain stores, and the mail-order business, facilitated by Rural Free Delivery.

After 45 years

After 45 years, the store was reopened—not as an agricultural village service center—but as part of a tourist attraction catering to those enthusiastic about our heritage. Today, the fixtures, containers, products, the empty tins, wooden boxes and other artifacts of the old country store are popular collectibles.

Above—Joel Miller's 1851 General Merchandising Store, Witmer, Penna., as it appears today—a century and a quarter later—as a restored attraction at the Folk Craft Museum of regional artifacts and Americana.

A Shenandoah Valley general store of the 1929 period when such enterprises were still floundering in isolated areas.

Cass, West Virginia, was established as a company town in 1900 when lumbering interests organized a major center to remove the hardwood from this extensive forest.

Cass Company Store in West Virginia, once claimed to be the largest company store in America.

The Cass Lumber Rail Road is now a state-owned historic site featuring scenic rides in trains pulled by original Shay engines.

Facing page

The drug department of the Cass Company Store. There was also a doctor available through the company to serve the workers.

Pocahontas Company Store

CASS, WEST VIRGINIA.

☛ A company store was established in 1900 when the Spruce Lumber Company acquired timber rights to nearly 200,000 acres on Cheat Mountain in West Virginia. In order to cut, haul and process the lumber, a lumber railroad had to be built, loggers recruited, mill workers hired and a community constructed.

The Pocahontas Supply Company was created to serve this new community. In its large building were three stories and the basement. Here a family could purchase all its needed objects—the store had a grocery, pharmacy, furniture store, hardware, clothing, jewelry and farm supply departments. It had a soda fountain too! Adjoining it was a slaughter house, butcher shop, and a coal and fuel depot.

Two wagons hauled supplies to residents. At night clerks loaded supplies on the lumber train to handle the needs of up to 1,200 loggers and track workers living on Cheat Mountain. Hay, feed and equipment for nearly 200 draft horses had to be hauled daily over the more than 100 miles of track.

The mountain logging communities were served exclusively by the store—Whittaker and Spruce. Spruce was the highest and coldest community in eastern America—once the temperature fell to 42° below zero! Residents had two means of transportation to the outside world—their own feet or the logging railroad.

The entire logging operation ended in 1960—the mills are closed and the towns on the mountain are extinct. But the railroad survives as a tourist attraction, along with the Company store.

Barter and Credit

☛ The storekeeper traded two ways—he exchanged goods with his local customers, and he bought and sold goods with wholesalers and commission merchants. When a family brought in butter or eggs for credit, he had to evaluate current market value and quality before entering the credit in his day book and the family account book. His major problem was that products varied in price through time and in quality, and he had to adjust credit accordingly.

As the original account record (above) shows, when the store at Mt. Crawford, Virginia, sold the butter it had taken in trade, the price received for it varied as much as 20% within the same shipment based on quality. So each transaction required specific attention rather than a one-price policy basis.

KISER & SHUTTERS,
DEALERS IN
GENERAL MERCHANDISE.
HIGHEST MARKET PRICE PAID FOR ALL KINDS OF COUNTRY PRODUCE.
Mt. Crawford, Va.,

JOHN S. MILLER. F. W. YAGER.

Account Sales of Butter

Sold for and on account of Messrs Kiser & Shutters

By MILLER & YAGER,

Nos. 911 Louisiana Avenue and 912 C Street, Washington, D. C.

Augst 20 — , 1875

4 Tubs Butter 712 27c 36 25 46 224	49	36
1 " do 46 264	11	96
	$61.	32
1 Tub Butter on hand		

Confusing Money

Inland merchants had unique problems—in very early times, they were faced with complex monetary conditions because both English and Spanish currency and local bank issues were used, all varying widely in value from place to place.

Weights and Measures

There were frustrating problems of weights and measures, often without adequate facilities. Buying in bulk they had to contend with shrinkage. Most merchants marked items as they were received, using a simple code such as the phrase CASH PROFIT which was converted mentally into numbers. If a label read SRI-$4.92, it would mean the article cost $3.69 (the 3 from the third letter in the code word—S, etc.), the mark-up was one-third, so the price was $4.92.

Country storekeepers in remote areas were usually important community figures, often among the limited few who could read and write. Being in touch with the major markets kept them well-informed. They had to keep accurate records because much of the business was on credit and, in the agricultural areas, customers often settled up only once a year—after the crops were in!

When a family member bought at the store, it was charged to the account whether 3 stoneware crocks, 8 yards of flannel, or a bag of salt—eventually paying in cash or trade goods. Eggs and butter were often brought in regularly to be credited to the account, but the trade goods extended over a wide range which could vary with the seasons.

What Prices!

The March 7, 1890 weekly broadside of Justice & Sandbern, Commission Merchants on Dock Street, Philadelphia, offers some idea of the range of goods they purchased from the General Stores. In addition to the price quotes on butter, eggs, poultry, sheep, vegetables and fresh fruit, some of the other items listed were:

Item	Price	Item	Price
Cow hides	2–5¢ lb.	Beeswax	21–26¢ lb.
Wool	23–33¢ lb.	Dried Fruit:	
Feathers:		Apples	3–4½¢ lb.
Goose	45–47¢ lb.	Peaches	4–10¢ lb.
Chicken	4–4½¢ lb.	Raspberries	21–22¢ lb.
Turkey	2½–3¢ lb.	Blackberries	4–4½¢ lb.
Duck	20–25¢ lb.	Cherries	9–11¢ lb.
Squab (alive)	30¢ pr.	Plums	7¢ lb.
Pigeon (alive)	40¢ pr.	Walnut kernels	9–12¢ lb.
Lard	6–7¢ lb.	Ginseng	$2.00–$2.75 lb.
Tallow	4–4½¢ lb.		

Other items bought included hides, chestnuts, and a family could earn considerable credit by shelling black walnuts, rendering lard from butchering, searching natural bee hives or keeping an aviary, drying feathers, gathering fruits. Gathering ginseng was a common activity in mountain areas, serving some folks as a "cash crop". Customers without an account at the general store were subject to the standard policy of "cash and carry"!

1929

It is here. So are we and So are you. Thanks be to a Kind Providence. Now permit us to serve you better for 1929. Your patronage will help us to do so.

E. G. HOOVER
56 N. C
HA

1929

If you followed us in our advertising during 1928 we hope we served you well. It will pay you to follow us closer in 1929. It will pay you better both in values and service.

Our Specials Friday & Saturday, Jan. 11 & 12

Urma Pancake Flour **9c**	Tis So Good Ketchup 8 oz. Bottle **9c**	Shredded Wheat **10c**
New Crop **Cal. Prunes** 40 - 50 Size 2 lbs for **23c**	Double Tipped Large Size Box **Matches** 3 for **11c**	
Spredit Margarine Was the First and the Best to sell for **25c** Continuous use in families five years, speaks for its quality.	For Body and Muscle Building Food eat Caskey's Old Home Potato Bread. Large loaf **10c**	Joy and Comfort After Drinking **Joy Cup Coffee** **50c**
High Grade Fine Tissue Paper 3 for **17c**	Octagon Soap Powder 2 for **13c**	Pink Salmon Tall can **17c**
Macaroni, Spaghetti or Noodles 3 for **20c**	New Process, Quick Cooking or regular style. Urma Oats **9c**	
Bee Brand Spices, Teas, Flavoring Extracts, under the Bee Brand, the best that money can buy. Packed by McCormick and Co. Known everywhere.	Sudden guests Always delighted with Sandwiches made from Majestic Sandwich Spread. Fine for a midnight lunch. **10c and 25c jars**	

To meet competition from rapidly developing chainstores, the General Store offered weekly specials promoted by handbills. Some wholesalers cooperated with special pricing and even printing the handbills. The specimen at left was supplied to the Hagerstown store by McCormick, of Baltimore, featuring its Bee-Brand label. Note the price—it was January 1929, but these reflect predepression price level. (Oh, For those Good Ole Days!)

Color Helps to Sell

☛ The development of color lithography brought with it a deluge of bright, colorful counter posters, signs and advertising cards. These were published by suppliers for use in stores to attract attention to their product.

On the facing page, the colorful advertising card of *Old Judge* was one of thousands of various tobacco advertising cards. The competition was keen between cigars and chewing tobacco and there were numerous brands of cigars prior to the popularity of cigarettes. *Ivory Salt* advertised *flavor* but *Morton* noted the annoying tendency for salt to lump and focused advertising on dryness—"it pours!"—and won the marketing battle.

Colorful cards featured scenes, florals, animals, patriotic themes, historic sites and comic situations. These were premiums inside the packaged goods or were made available by writing the manufacturer. Commonly collected, they were placed in scrapbooks, albums or given to loved ones as small gifts.

These chromoliths are among the contemporary collectibles and are found at many flea markets and antique shops today.

Towles Syrup came in a tin can shaped like a log cabin; a special counter top display was offered briefly to promote the wholesale purchase of the product.

Blacking for the care of leather was a standard product in almost every store.

Printing on tin, porcelain and cardboard. The tin and porcelain were heavy duty ads which could be placed outside and withstand the weather. The advertising card at lower right dates to 1886 when the New York Condensed Milk Company sold the Gail Borden Brand Condensed milk—this later became the Borden Company.

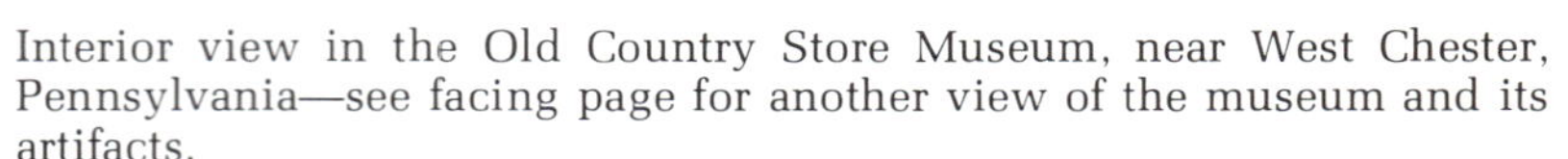

Interior view in the Old Country Store Museum, near West Chester, Pennsylvania—see facing page for another view of the museum and its artifacts.

Typical account books used for family records of purchases made at the local store.

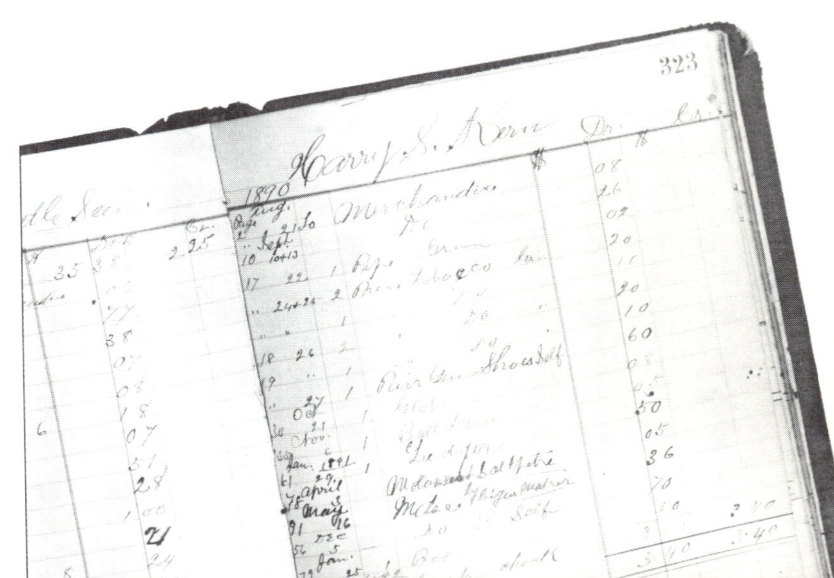

Farm Supplies

☛ Every store stocked farm supplies, seeds being a major item, but those used by the farmer were merchandised differently than those used by the womenfolk for their kitchen and flower gardens. By 1860 garden seeds were attractively packaged in colorful lithographed packets to stimulate interest in gardening and visual hope for success and reward for the energy expended.

Early Seed Distributor

One seed producer and distributor was Charles W. Briggs who established his company in Rochester, N.Y., in 1845. His business flourished, and after his brother became a partner in 1861, they operated as Briggs Brothers.

Briggs Brothers featured a handy combination shipping container and counter display containing a wide variety of garden seeds and variants for the most popular vegetables. Such merchandising containers stimulated wholesale sales to general stores because of their convenience for the operator. They went out of business in 1919.

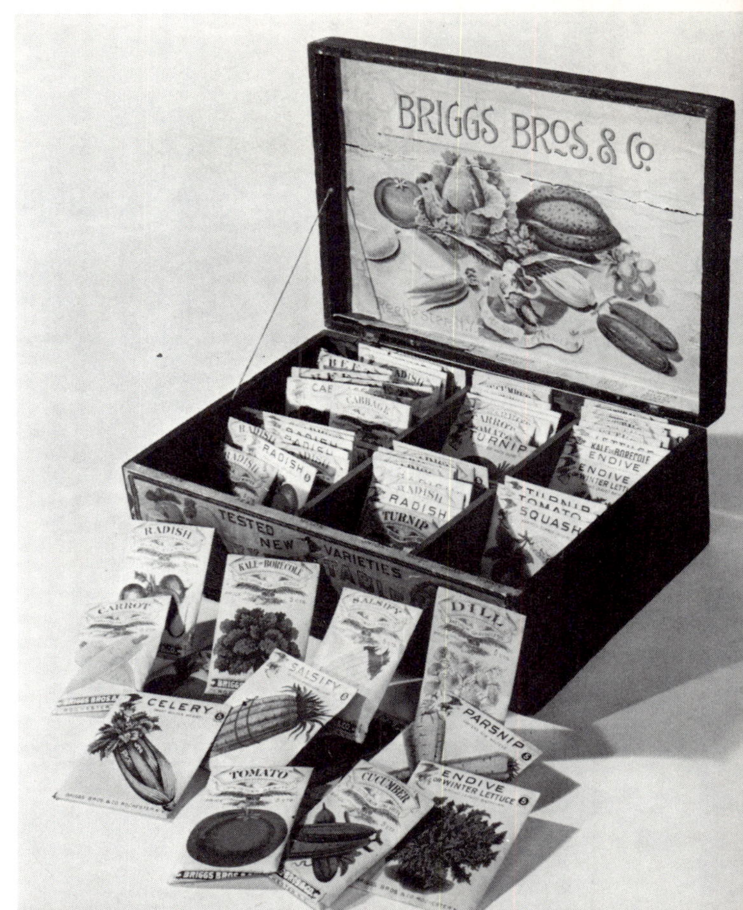

Display cases for counter or floor use were made available by producers to assure that the product would find its place in the confusion. *Diamond Dye* or its direct competitor *Putnam* had cases in most stores; ribbons and silk thread were safely displayed by Belding walnut cabinets. Combs were placed in rotating units—complete with mirrors!

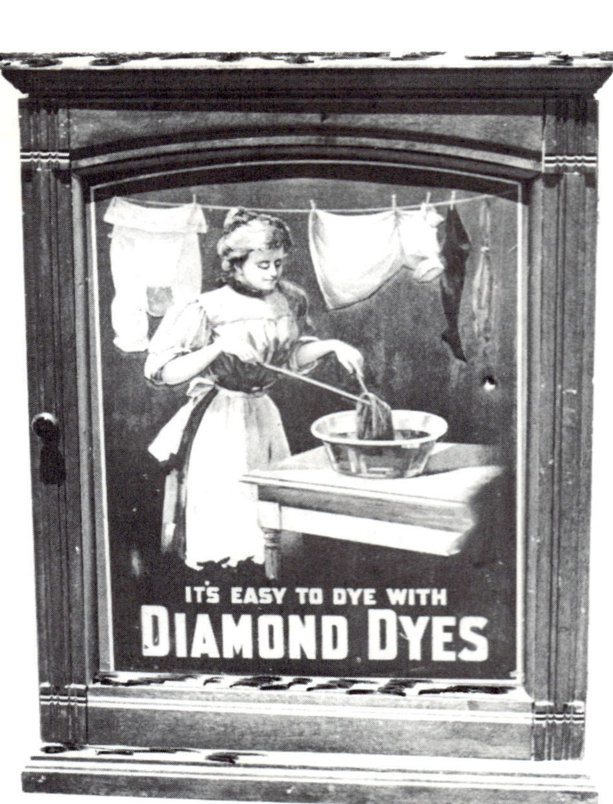

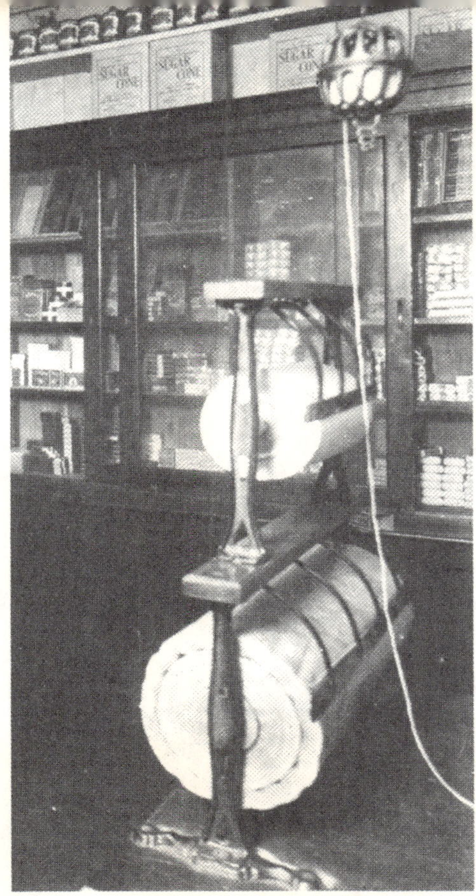

Clark's O. N. T. (Our New Thread) was usually found in one of several types of counter cases; the tin A & P dispenser was a permanent fixture in its red & gold color; the wooden Jersey Coffee floor display was also the shipping case for 120 one-pound bags of coffee beans. The counter wrapping paper unit and string arrangement was a busy center prior to the development of the paper bag.

"We've got it, if we can find it" was a common statement of the country store clerks. Products ranged from rat traps to fly paper—from horse liniment to household linens—creating a clutter of disorganization and chaos.

SUPERIOR
BISCUIT
CRACKER
FINE CRACKERS
COFFEE
PEANUT BUTTER
Manru Coffee
TUTTLE & SPICE
OLD GLORY
MOCHA & JAVA
FLAVORED
6 BLOCKS FOR 5 CENTS
USE
MAGIC BLUE
6 BLOCKS FOR 5 CENTS

Display of products in the Tuttle and Spice
1880 General Store, near New Market, Virginia.

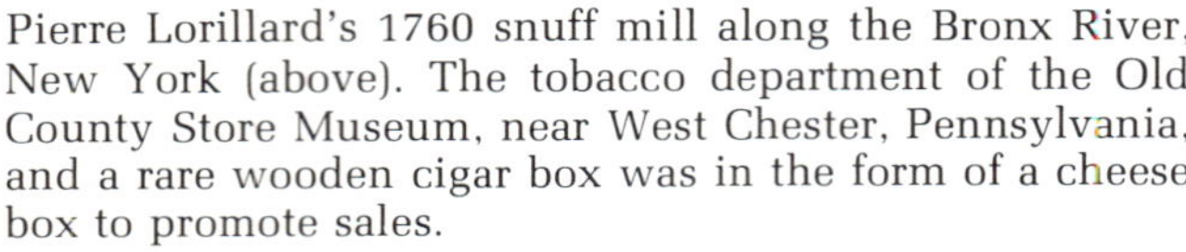

Pierre Lorillard's 1760 snuff mill along the Bronx River, New York (above). The tobacco department of the Old County Store Museum, near West Chester, Pennsylvania, and a rare wooden cigar box was in the form of a cheese box to promote sales.

Tobacco and The General Store

☛ Collectors of Americana have had an unusual interest in tobacco products and the associated artifacts such as tin plug tags, cigar bands, cigar box labels, tobacco tins, collector cards and the carved wooden images of Indians and others. The history of tobacco in America is fascinating because by 1900 there was a multitude of tobacco brands on the market—12,600 chewing brands, 7,000 smoking brands and 2,100 cigar and cigarette brands.

Yes, They Stocked It!

The general store stocked tobacco products in several forms including snuff and plug tobacco. (The early peddler usually sold "pigtail" which was a coiled, rope-like form sold by the foot or yard at around 3¢ per foot.)

Store owners were induced to stock brands by promotional gimmicks such as counter displays, cigar cutters, wall lighters, and printed matter ranging from the sport and historic figures of the time to leg art. The consumer was rewarded with picture cards and premiums for prizes.

Wooden Indians and Such

The American Indian, long associated with tobacco, became the emblem of the cigar store and tobacco specialty shop. These carved wooden figures are now rare American folk art, yet they were common in city stores prior to the Civil War. There are Scotsmen holding snuff, Punch, Pioneers, Sailors, and a Baseball player figure—but nine out of

ten were Indians holding cigars or a leaf of tobacco. Small wooden images were made for counter tops.

Later the wooden images were vandalized or stolen, so they were placed on stands with rollers and taken in at night. Around 1900 laws against obstruction of the sidewalks reduced their use.

Snuff and Plug

The early use of tobacco is reflected in the fact that Pierre Lorillard built a snuff mill in New York in 1760 and a large manufacturing plant in Jersey City where plug tobacco was produced. Lorillard branded each plug with colorful printed tin tags pronged in. These trademarks were used as premiums redeemable for prizes. Bulk smoking tobacco was also competitive, with hundreds of variant brands.

During the eighteenth century tobacco leaves were fashioned into a twisted form—yards long. This rope-twist was sliced into shavings and used in a pipe—ground for snuff, cut into bite-size for chewing.

As chewing became popular, plug tobacco was created by using pressure devices to compress the tobacco into a cake. Plugs weighed one pound and measured 3″ x 16″. A clerk sliced it into five or six "cuts" which sold at 10¢. Available either natural or flavored, the most common flavor being licorice, but nutmeg, rum, cinnamon, maple sugar and prune were also used to capture those with a sweet tooth.

The Early 19th Century Cigar

The rise of the cigar came during the first half of the nineteenth century. Three clear levels emerged—the expensive Havana's, the cheaper blended domestic cigars, and the cheap cheroot, stogie and toby.

The *cheroot* was an unblended and untapered roll of tobacco. The *stogie* was a foot-long domestic tobacco and the *toby* was made the same as the

Plug tobacco cutter from Folk Craft Musuem, Witmer, Pa.

stogie but was sweetened with molasses, and it could be chewed or smoked.

By 1890 cigars were highly competitive: hundreds of "twofers" (two-for-five-cents) were on the market. Cigars featured a band with colorful embossed trademark and box labels. These are both among today's collectibles.

Cigarettes

Cigarettes were not popular at the turn of the century—in 1880, 60% of all cigarettes were manufactured in New York City. In 1910 they were still primarily an urban smoke, featuring a Turkish association as evidenced by names such as: *Turkish Trophies*, *Egyptian Deities*, *Murad*, *Helmar* and the later number one brand, *Camels*.

By the time cigarettes dominated the tobacco counters, the general stores were fading from the scene almost as rapidly as the cigar store Indian!

Tobacco case intact from an Ellicotville, New York, store as it was in 1900, now on display on Tuttle and Spice 1880 General Store.

A popular card advertisement for Jackson's Best Navy Plug tobacco—turn it upside down and the Jackass becomes a man!

Two shelves of tobacco at the Witmer Country Store; and shelves from the tobacco section of Tuttle and Spice 1880 General Store—chewing and smoking were obviously common habits!

Baked beans with salt pork and molasses were a regular New England dish not generally popular elsewhere. But in 1861 Gilbert VanCamp, an Indianapolis grocer, combined navy beans with tomato sauce (some folks claim it resulted from a fire in the warehouse!) and the new product became nationally popular. Almost every major packer took up the combination, making it one of the nation's most popular canned good items.

Early Canned Goods

From early times primitive people sought to preserve foods for future use. Canning revolutionized the preserving process and permitted the use of many more types of goods including ready-to-serve varieties.

The term *can* is derived from the Greek *cane* from which baskets were made. Baskets were originally used as containers called canisters. Our early tin containers were referred to as canisters, later shortened to *tins* or *tin cans*.

William Underwood established a cannery in Boston in 1821. Noting the resistance to canned products because of the metallic smell, he masked that with spices in the meats called "deviled"—thus pioneering in preserving. Another pioneer was Gail Borden who condensed milk products and *Bordens* became a major American corporation.

Scene in the Old Country Store Museum (above). When spring came some ladies bought new bonnets, men most likely dressed up the old straw hat with a coat of coloring!

Fresh roasted peanuts were always a popular treat—this old machine is on display at Tuttle and Spice 1880 General Store.

Anyone Can Make Soap

☛ Several years ago, a 92-year-old mountain woman was asked how she made soap. She retorted, "Anybody kin make it—don't a growed man like you know how?" Soapmaking was a common household task in early times. Fats were saved from trimmings at the fall butchering and wood ashes from the stove made lye. Yet it was virtually an impossible task in emerging towns and cities.

Three Made Fortunes

In spite of the fact that "anyone could make soap", very few developed soap-making into major corpora-

The store poster was originally drawn by Henry Furness for Pear's Soap and was used in the London magazine *Punch*. In America it was adopted by *Kirk & Co.* and used for several brands of "White Russian" and "Kirk's Flake." The company later was merged with Procter and Gamble.

Ivory soap became the most widely known soap in America, being the first to be advertised nationally with such honored slogans as "99 44/100% Pure" and "It Floats!"

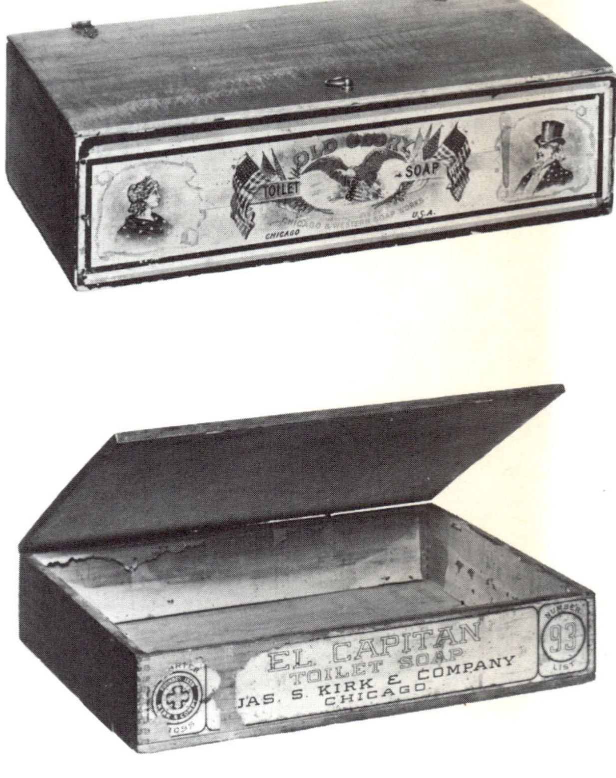

tions! Three Americans made fortunes—William Colgate, James Gamble and Benjamin T. Babbitt.

Colgate was born on a Maryland farm in 1783 and took to making soap in Baltimore and later in New York with John Sliddell, a tallow chandler. In 1806, he set up his own business and became so successful that Madison University changed its name to Colgate University in his honor and his corporation still flourishes today.

Gamble made soap in Cincinnati, selling the product locally until 1836 when William Procter entered into a joint distribution venture to sell candles. In the resulting partnership, candles were featured with soap secondary. When the demand for candles decreased, the partners concentrated on soap and in 1879 Procter and Gamble introduced the first brand name soap—*Ivory*, which was made in large cakes with notches so it could be broken in half.

Babbitt started making soap in 1843 near Oneida, New York, and peddled it along with baking powder and saleratus throughout western New York. He moved his enterprise to New York City where he introduced small bars of wrapped soap. (The common practice at the time was to produce large loaf sizes. The store clerk cut a slice from the loaf to suit his customer.)

Babbitt faced severe sales resistance to wrapped soap in small bars, so in 1851, he placed a premium offer in each wrapper. The public responded to "something-for-nothing", and his sales soared!

A Pear's soap original poster; the soap and laundry goods available in the past as displayed in the Tuttle and Spice 1880 General Store exhibit (below).

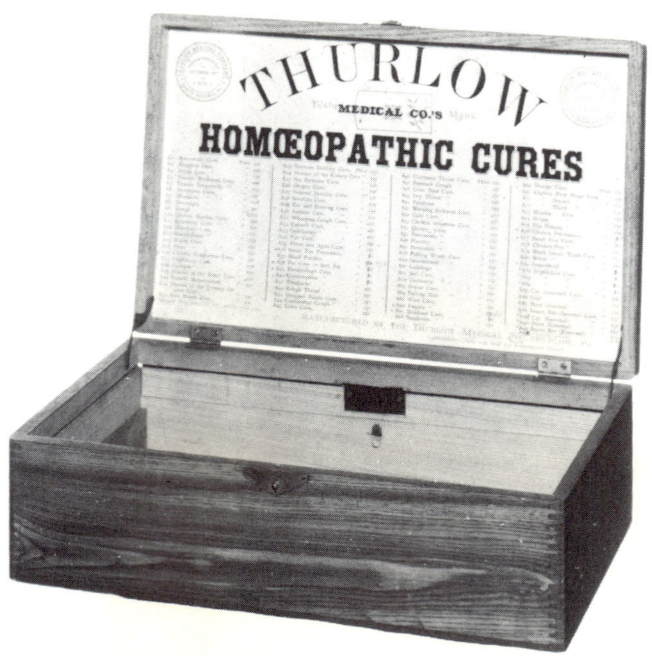

"Doctoring" and the General Store

☛ The general store persisted in agrarian areas with sparse populations at a distance from towns and cities. Such farmsteads were semi-isolated and the need to seek self-sufficiency was evident. There were few professional services so the households continued to do much of their own building, sewing, baking, blacksmithing, and even doctoring.

Patent Medicines on the Shelves

The hard work led to plenty of aches and pains—thus a demand for convenient ready-made remedies for common ailments. This demand was met by more than 3,000 medicine makers marketing more than 10,000 brand names or labels. Sales reached around 300,000,000 bottles or containers a year. It was a *big* business!

Early Advertising

Ready-made (so-called Patent) Medicines became available in most country stores through demand created by local newspaper advertisements. It has

Thurlow medicine counter display case (above) and the advertising poster of Dr. A. C. Daniels along with select specimens of his product (below).

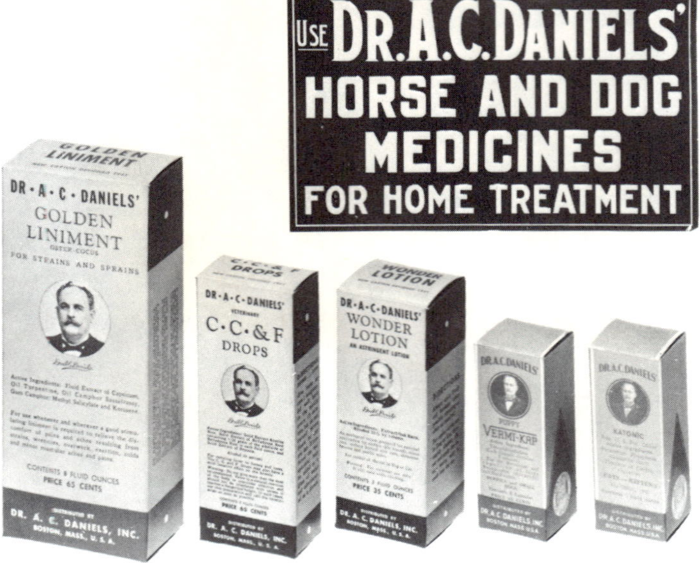

The shipping box for a product called "The People's Remedy" made by the Gilbert Brothers of Baltimore, and the dozen unit shipping box for Yager's liniment good for man or beast!

been estimated that 20,000 local papers ran patent medicine ads at the turn of the century. It is noted that *Lydia E. Pinkham's Vegetable Compound* was promoted with $50,000,000 in newspaper and magazine advertising! Little wonder the local cross-roads store stocked shelves of various medicines.

Traveling "Doctors"

There were also itinerant medicine-men dispensing their medicines to patients and selling quantities to general stores at wholesale prices.

A Dr. Sweet of Connecticut traveled widely. His medical consultations were free but the medicines he dispensed were not. Dr. Fishblatt of Iowa traveled three weeks each month examining and dispensing medicine over an extended region. Dr. J. M. Howard of North Carolina offered a free show before he sold his medicines. Billed as the "Great Southern Ventriloquist and Medical Consultant", he too had his own medical labels.

When the itinerant medicine-man was not available, people naturally turned to the country store for patent cures.

An original colorful counter poster advertising Barker's Liniment. These were 11" x 14" and given to country stores to promote the product.

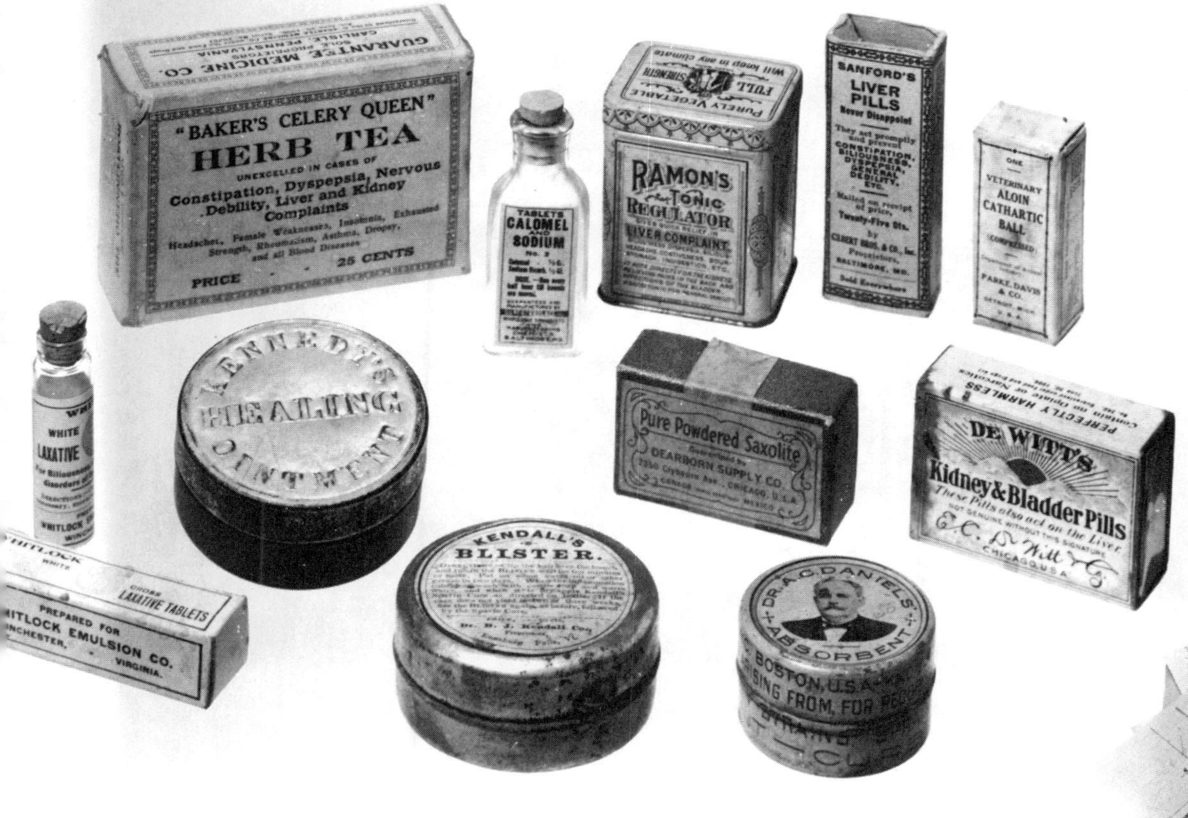

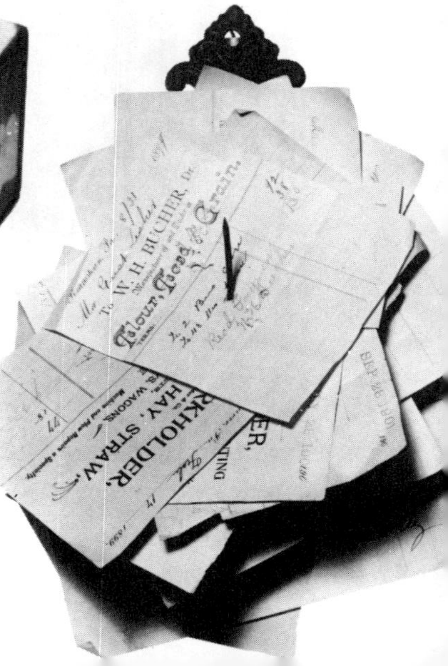

In the early times, and it seems to be recurring through to the present, cash was always in short supply and bills accumulated faster than the payment of same!

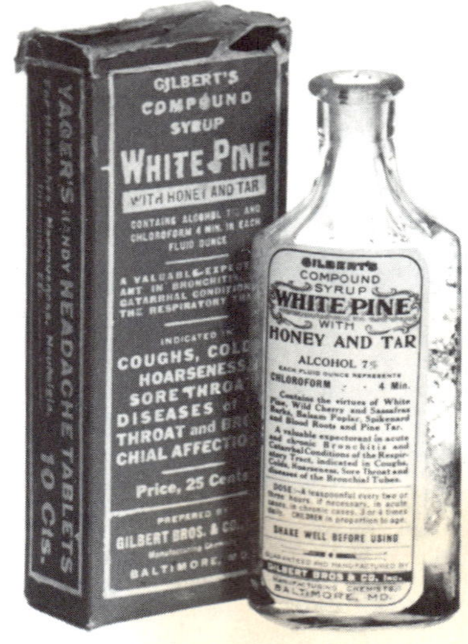
GILBERT'S COMPOUND SYRUP
WHITE PINE
WITH HONEY AND TAR
COUGHS, COLDS
HOARSENESS
SORE THROAT
Price, 25 Cents
GILBERT BROS. & CO.
BALTIMORE, MD.
YAGER'S HANDY HEADACHE TABLETS
10 Cts.
GILBERT'S COMPOUND SYRUP
WHITE PINE
WITH HONEY AND TAR
ALCOHOL 7%
SHAKE WELL BEFORE USING
BALTIMORE, MD.

SHILOH
COUGHS
COLDS
SPASMODIC CROUP
HOARSENESS
WHOOPING COUGH
SORE THROAT
S. C. WELLS & CO.
SHILOH
FOR
COUGHS, ETC.
S. C. WELLS & CO.

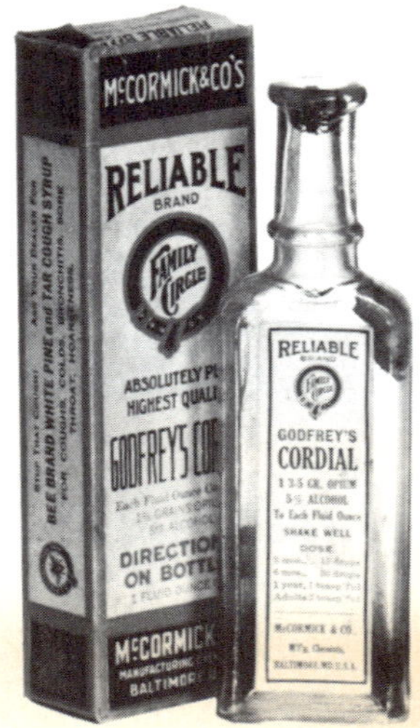
McCORMICK&CO'S
RELIABLE
BRAND
FAMILY CIRCLE
ABSOLUTELY PURE
HIGHEST QUALITY
GODFREY'S CORDIAL
DIRECTIONS
ON BOTTLE
McCORMICK & CO.
BALTIMORE
RELIABLE
GODFREY'S
CORDIAL
SHAKE WELL
DOSE
McCORMICK & CO.

FOLEY'S
HONEY AND TAR
COMPOUND
Throat, Chest and Lungs
FOLEY & CO.
CHICAGO, ILL.
FOLEY'S
HONEY AND TAR
COMPOUND
FOLEY & CO.

Kodol
TRADE MARK
DYSPEPSIA
CURE
The New Digestant
Combined with
PEPSIN,
DYSPEPSIA
PREPARED ONLY BY
E.C.DeWITT & CO.
CHICAGO, U.S.A.
PRICE 50 CENTS
KODOL DYSPEPSIA CURE
KODOL DYSPEPSIA CURE.

Rydale's
WHITE PINE
Compound
RYDALE REMEDY CO.
Newport News
Price,
Rydale's
WHITE PINE
Compound.
RYDALE REMEDY CO.
NEWPORT NEWS, VA.
PRICE 25 CENTS

ORINO
LAXATIVE
FRUIT SYRUP
PLEASANT TO TAKE
Cures
Liver Trouble
PREPARED ONLY BY
FOLEY & CO.
CHICAGO, ILL.
Price 50 Cents
ORINO
LAXATIVE
FRUIT SYRUP
COMPOUND
Pleasant to take
THE NEW LAXATIVE
FOLEY & CO.
CHICAGO

Rydale's
TRADE MARK
Emulsified Oil
LINIMENT
FOR MAN AND BEAST
PRICE, 25 CENTS
PREPARED BY THE
Rydale Remedy Co.
NEWPORT NEWS, VA.
Rydale's
Emulsified Oil
LINIMENT
IMPORTANT
Rydale Remedy Co.
Newport News, Virginia.

SELSMONIA
A Liquid
Preparation
for Coughs, Colds,
Influenza and
Pneumonia
DIRECTIONS
THE SELSMO COMPANY
SELSMONIA
A Liquid
Preparation for
Coughs, Colds,
Influenza, Pneumonia
CHILDREN:
PREPARED BY
THE SELSMO COMPANY

GILBERT'S
No. 10
BENZINE
REDISTILLED
AND
PERFUMED
FOR ERASING
GREASE SPOTS
FROM CLOTHING
KID GLOVES, ETC.
PREPARED BY
GILBERT BROS.
BALTIMORE
GILBERT'S
REDISTILLED
AND
PERFUMED
BENZINE
BALTIMORE

DR. W. B. CALDWELL'S
SYRUP PEPSIN
Herb Laxative Compound
FOR CONSTIPATION
PEPSIN SYRUP CO.
MONTICELLO, ILL.
DR. W. B. CALDWELL'S
SYRUP PEPSIN
LAXATIVE SENNA COMPOUND
FOR CONSTIPATION
PEPSIN SYRUP COMPANY
MONTICELLO, ILL.

BYRD'S
ORIENTAL
BALM
COMPOUND
BYRD'S
Oriental Balm
A REMEDY
PRICE 50 CENTS
BLUE RIDGE CHEMICAL CORPORATION
ROCKY MOUNT
COMPOUND
FOR INTERNAL AND
EXTERNAL USE
PRICE 50 CENTS
BLUE RIDGE
CHEMICAL CORP.
ROCKY MOUNT, VIRGINIA

HITE'S PAIN CURE FROM S. P. HITE STAUNTON, VA.

Hite's Pain Cure
Contains
ALCOHOL 65%
Oils, Gums, Etc.
Guaranteed under the FOOD and DRUGS ACT, June 30, 1906. No. 2384

The lettering above is from a stencil used by Hite's for shipping cartons and crates. The company later moved to Roanoke, Virginia.

The pain cure contained 65% alcohol. Its main contribution was that it came in very small bottles!

The Barker's Almanac was given away to country stores with the wholesale purchase of the products. The store in turn sold it or gave it away at the operator's discretion. Book marks were given with the purchase of Mennen's powders and cosmetics.

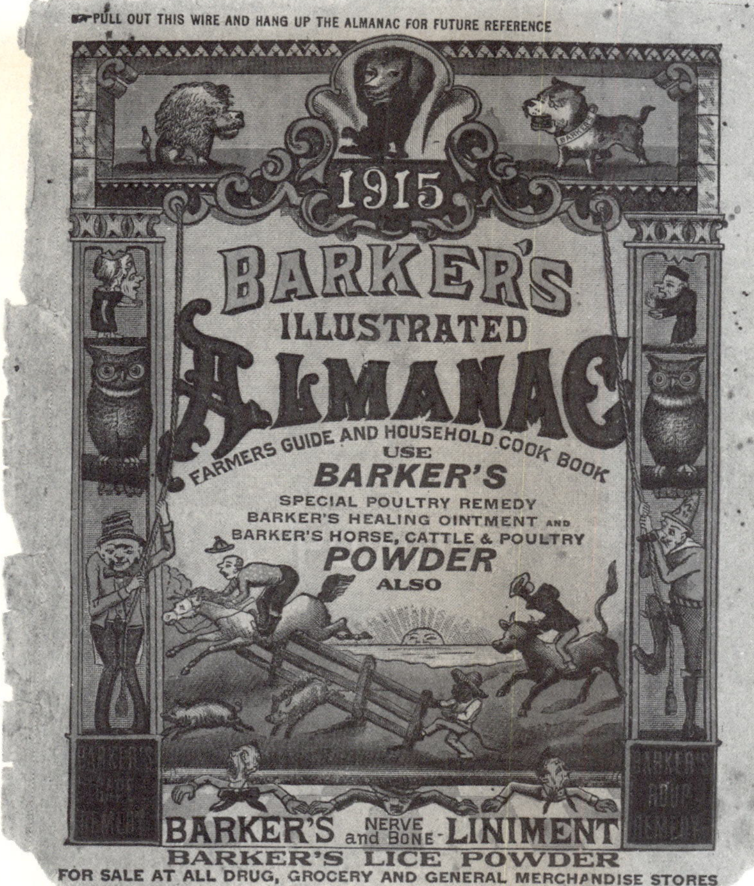

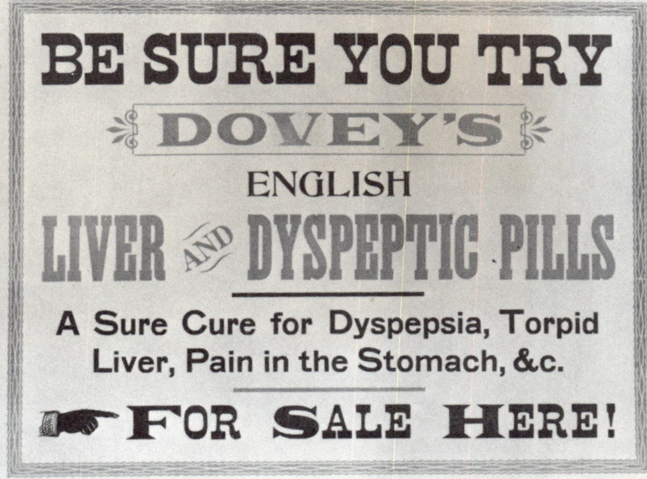

A scene in Old Village Store, Bird-in-Hand, Pa.

The kerosene pump at the Folk Craft Museum & Country Store, Witmer, Pa. dispensed the liquid "coal oil" into small containers from a large reserve tank in the cellar.

Then There Were Bags

Grain, flour and other loose commodities were shipped in cotton sacks or wooden hogsheads until the Civil War created a shortage of cotton. This stimulated a demand for a substitute and the paper bag was developed

In 1852 the first patent was issued on a paper bag making machine, but little interest in it was shown until 1865 when another was developed which cut paper, folded it and dampened the fold points with a flour paste and sealed it into a bag.

In 1869 the Union Company combined the best aspects of all previously developed machines finally creating an acceptable product. The paper bags were first used by wholesalers in shipping and pre-packaging. Eventually, stores used them as a relief from the tedious wrapping and tying of each commodity. Still later, the large shopping bags came along to replace market baskets. Bag racks and holders became commonplace in busy general stores to make various size bags readily available to operator and customer alike.